Start the Car

by Lelia Mander

BLACKBIRCH
PRESS

San Diego • Detroit • New York • San Francisco • Cleveland • New Haven, Conn. • Waterville, Maine • London • Munich

© 2004 by Blackbirch Press™. Blackbirch Press™ is an imprint of The Gale Group, Inc., a division of Thomson Learning, Inc.

Blackbirch Press™ and Thomson Learning™ are trademarks used herein under license.

For more information, contact
The Gale Group, Inc.
27500 Drake Rd.
Farmington Hills, MI 48331-3535
Or you can visit our Internet site at http://www.gale.com

ALL RIGHTS RESERVED
No part of this work covered by the copyright hereon may be reproduced or used in any form or by any means—graphic, electronic, or mechanical, including photocopying, recording, taping, Web distribution or information storage retrieval systems—without the written permission of the copyright owner.

Every effort has been made to trace the owners of copyrighted material.

Photo Credits: see page 47.

LIBRARY OF CONGRESS CATALOGING-IN-PUBLICATION DATA

Mander, Lelia
 Start the car / By Lelia Mander.
  v. cm. — (Step back science series)
 Includes bibliographical references and index.
 Contents: Start the car with a key — Parts of a car — Engine runs on explosions — Gas comes from oil refinery — Heat and pressure turn organisms into oil.
  ISBN 1-56711-678-7 (hardback : alk. paper)
  1. Automobiles—Juvenile literature. 2. Petroleum--Juvenile literature. [1. Automobiles.] I. Title. II. Series.

TL147 .B25 2003
629.222—dc21
                                                                2002153746

Printed in the United States
10 9 8 7 6 5 4 3 2 1

Contents

Start the Car

Start the Car with a Key 6
What makes a car start?

Key Turns Ignition On 8
How does turning a key inside a car make the car start?

Electricity Flows from Battery to Engine 10
Where does the electric current come from?

▶ **SIDE STEP: Charge It!** 12
Can a car run just by using the battery?

Engine Runs on Explosions 14
Why is a spark needed to start the engine?

Mixture of Gas and Air Explodes and Moves Pistons 16
What does the spark set fire to?

▶ **SIDE STEP: Timing Is Everything** 18
If engines are exploding all the time, why do they seem to run so smoothly?

Driver Hits Gas Pedal to Squirt Gas in Engine 20
Where does the gas come from?

Gas Added to Car at Gas Station 22
How did the gas get into the gas tank?

▶ **SIDE STEP: Making the Grade** 24
Are there different kinds of gas?

Trucks Carry Gas to Underground Tanks 26
How does the gas get to the gas station?

Gas Comes from Oil Refinery 28
Where do the trucks pick up the gas?

Gas Comes from Crude Oil at Refinery 30
How is gas made at the refinery?

▶ **SIDE STEP: Waste Not, Want Not** 32
What else is produced at the refinery?

Crude Oil Drilled from Large Underground Reserves 34
Where does crude oil come from?

Heat and Pressure Turn Organisms into Oil 36
What is crude oil made of?

Fossils Buried Under Sea Hundreds of Millions of Years Ago..... 38
How did the fossils get deep underground?

The Big Picture 40
Learn how a car starts and what makes it go.

Facts and Figures 42

Wonders and Words 44
Questions and Answers

Glossary 45

Index 46

For More Information 48

How to Use This Book

Each Step Back Science book traces the path of a science-based act backwards, from its result to its beginning.

Each double-page spread like the ones below explains one step in the process.

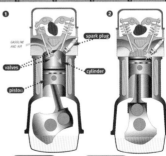

A time line along the top describes all the steps in the process. A marker indicates where each spread is in the process.

A question ends each spread and is repeated as the title of the next spread.

A short description gives a quick answer to the question asked at the end of the previous step.

Sidebars show interesting related information.

Side Step spreads, like the one below, offer separate but related information.

The Big Picture, on pages 40–41, shows you the entire process at a glance.

| Start the Car with a Key | Key Turns Ignition On | Electricity Flows from Battery to Engine | Engine Runs on Explosions | Mixture of Gas and Air Explodes and Moves Pistons | Driver Hits Gas Pedal to Squirt Gas in Engine | Gas Added to Car at Gas Station |

Start the Car with a Key

What makes a car start?

Cars are everywhere. Most car owners in the United States use their car several times a week, though many drive every single day. People need cars to get from one place to another and to carry things like groceries and supplies.

What makes a car start? The simple answer is that the driver puts a key into the ignition and twists. In fact, the answer is a lot more complicated than that. Under the hood of a car there is a maze of different parts and mechanisms that all must work together to get things started. The process of starting the car only begins when the key is put in the ignition.

So how does turning a key inside a car make the car start?

battery
The battery is the source of a car's electrical power. A car needs electricity to start and to power lights, heating, and other electrical functions.

| Trucks Carry Gas to Underground Tanks | Gas Comes from Oil Refinery | Gas Comes from Crude Oil at Refinery | Crude Oil Drilled from Large Underground Reserves | Heat and Pressure Turn Organisms into Oil | Fossils Buried Under Sea Hundreds of Millions of Years Ago |

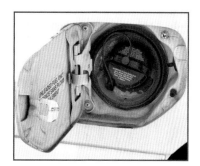

fuel tank
The gas needed to power the engine is stored in the fuel tank. Cars will not run unless they have fuel. For most modern cars, fuel means gasoline, or gas for short.

engine
The engine is the heart of the car and the source of its power. It converts gas to energy, which helps enable the car to start.

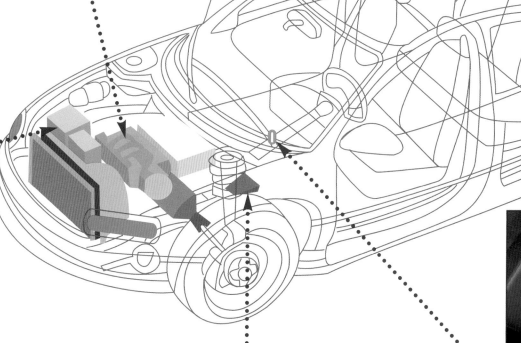

ignition
The ignition turns the car on and sends electrical power to the engine.

accelerator
The accelerator is the driver's connection to the fuel tank. The driver presses on this pedal to send fuel to the engine, which increases the car's speed.

Start the Car   7

Key Turns Ignition On

How does turning a key inside a car make the car start?

The key connects to the car's on/off switch, so a turn of the key switches the car on and off. For security reasons, each car comes with a uniquely shaped key, which operates its own ignition switch. Car theft would be too easy if cars started with a universal key, or with the push of a button.

Just as a switch must be flipped to turn on a light, a switch starts the process of turning on a car's engine. Electricity that moves from one place to another is a current. A car's ignition switch sends an electric current to two places. One is an electric motor called the starter. This motor makes the main part of the engine start to turn around. The car engine needs this first push before it can run. Electric current is also sent to an engine part called the coil.

But where does the electric current come from?

| Trucks Carry Gas to Underground Tanks | Gas Comes from Oil Refinery | Gas Comes from Crude Oil at Refinery | Crude Oil Drilled from Large Underground Reserves | Heat and Pressure Turn Organisms into Oil | Fossils Buried Under Sea Hundreds of Millions of Years Ago |

Get It While It's Hot

Experienced thieves know a way around car keys to start and steal cars. After they break into a car, thieves may pull out the wires in the car's ignition. By tampering with these wires and making the right connections between them, a thief could produce the electric current needed to make the engine run. This action is called hot-wiring because it creates a spark between the wires to warm up and start the car.

Recently, carmakers have made it harder to hot-wire a car. Many car keys now come with a tiny resistor embedded into them. A resistor is a device used to control currents in an electric circuit by providing resistance. The resistor in the key and ignition are designed together so that without a particular resistor, a car will not start.

| Start the Car with a Key | Key Turns Ignition On | **Electricity Flows from Battery to Engine** | Engine Runs on Explosions | Mixture of Gas and Air Explodes and Moves Pistons | Driver Hits Gas Pedal to Squirt Gas in Engine | Gas Added to Car at Gas Station |

Electricity Flows from Battery to Engine

Where does the electric current come from?

The ignition draws the electric current from the battery, a large box under the hood near the engine. The battery stores the electrical power necessary to start the engine and to work other electrical systems, such as lights, heating, and windshield wipers. A pair of cables connects the battery to other parts of the car.

For the car's engine to operate, fuel must burn. To begin the burning process, the battery sends the engine an electrical current by way of the ignition switch. However, the battery supplies only 12 volts of electricity. The engine needs a much larger current—14,000 volts—before its fuel can ignite.

The current gets that extra jolt of voltage through the coil, a cylinder-shaped engine part made of two wound coils of copper wire wrapped around a magnetic core. As the electricity travels from the outer coil to the inner coil, its voltage increases. This higher voltage is enough to create a spark.

But why is a spark needed to start the engine?

| Trucks Carry Gas to Underground Tanks | Gas Comes from Oil Refinery | Gas Comes from Crude Oil at Refinery | Crude Oil Drilled from Large Underground Reserves | Heat and Pressure Turn Organisms into Oil | Fossils Buried Under Sea Hundreds of Millions of Years Ago |

▲ A car's battery ignites its engine and powers its many electrical systems.

▲ Early cars had to be started using muscle power alone.

Real Muscle Cars

The earliest cars did not have ignition switches or starter motors. People had to use a lot of their own muscle power to get their cars started. The cars came with a hand crank that was connected to the engine. Drivers would turn this crank hard and fast to create the spark needed to start the engine.

Start the Car 11

Side Step

CHARGE IT!

Can a car run just by using the battery?

Fully electric cars do exist. Yet a battery alone does not have enough electrical power to start the engine of a gas-powered car, which is the most common kind on the road today.

Electric cars use electric motors powered by batteries that are much

bigger and heavier than standard car batteries. These batteries must be recharged about every 50 miles, so electric cars are not practical for long journeys. This could change with the recent development of fuel cells. These are energy-storing units similar to batteries, but they recharge themselves with hydrogen and oxygen instead of electricity. Fuel cells are expected to be included in newly manufactured cars by 2005.

Because they do not burn gas, electric cars do not create as much pollution as gas-powered cars do. They also use energy more efficiently and make less noise. For these reasons, many car industry insiders predict that electric cars will completely replace gas-powered cars by the year 2017.

◀ Electric cars are quieter and more fuel-efficient than gas-powered ones.

▲ Hybrid cars have both gas-powered and electric motors.

Have It Both Ways

Some cars use both gas and electricity to run. The cars, called hybrids, have a gas engine as well as an electric motor. The driver may, for example, use the gas engine for long journeys and then switch to the electric motor for small drives. Hybrid cars include a generator that can recharge the battery for the electric motor as the car moves.

| Start the Car with a Key | Key Turns Ignition On | Electricity Flows from Battery to Engine | **Engine Runs on Explosions** | Mixture of Gas and Air Explodes and Moves Pistons | Driver Hits Gas Pedal to Squirt Gas in Engine | Gas Added to Car at Gas Station |

Engine Runs on Explosions

Why is a spark needed to start the engine?

A spark sets off an explosion in the engine. This explosion is the first in a series of small explosions—hundreds of them a minute—necessary to make the car run.

A car engine contains a series of rods, called pistons, that move rapidly up and down when the engine runs. Each piston is inside a cylinder. Screwed into the top of each cylinder is a spark plug, which receives the electric current to make a spark. The spark creates an explosion that forces the pistons to move up and down in the cylinders.

The pistons are connected to a long rod called the crankshaft. As the pistons move, they turn the crankshaft. This action turns the wheels and makes the car move.

The explosions that make the car run are powerful releases of heat and energy. The initial spark sets them on fire, but the spark needs something to burn in order to make the explosion.

So what does the spark set fire to?

| Trucks Carry Gas to Underground Tanks | Gas Comes from Oil Refinery | Gas Comes from Crude Oil at Refinery | Crude Oil Drilled from Large Underground Reserves | Heat and Pressure Turn Organisms into Oil | Fossils Buried Under Sea Hundreds of Millions of Years Ago |

Sparking New Spark Plugs

A car engine cannot run without spark plugs. These metal devices are less than 2 inches long. Yet the electrical current that runs through the car must pass through them in order to produce an explosion inside each cylinder. Spark plugs do their job hundreds of times a minute in temperatures as high as several thousand degrees Fahrenheit. After a while, spark plugs wear out and must be replaced for a car to run properly.

Start the Car 15

| Start the Car with a Key | Key Turns Ignition On | Electricity Flows from Battery to Engine | Engine Runs on Explosions | **Mixture of Gas and Air Explodes and Moves Pistons** | Driver Hits Gas Pedal to Squirt Gas in Engine | Gas Added to Car at Gas Station |

Mixture of Gas and Air Explodes and Moves Pistons

What does the spark set fire to?

The spark sets fire to a mixture of gas and air. The mixture is inside each cylinder and is highly explosive, especially when the piston comes up and compresses it in the top of the cylinder. When the compression occurs, a powerful explosion goes off. The blast forces the piston down again.

After the explosion, the piston comes back up. At the top of the cylinder, a valve opens and releases the burnt fuel, or exhaust. As the piston moves down again, another valve opens and sucks in a new mixture of gas and air for the next explosion.

So where does the gas come from?

▲ *Because they are low to the ground, race cars are less affected by wind resistance.*

Tear Up the Road

The world's fastest cars are designed to be low to the ground to reduce wind resistance. Nothing helps cars achieve this better than airfoils. These horizontal panels at the front and back of a car enable wind to push the car close to the ground as it moves. This action also increases friction on the tires, which improves the car's grip on the road.

Besides airfoils, race cars have super powerful engines. In them, ten cylinders work together to turn the crankshaft much faster than ordinary car engines with four or six cylinders.

Trucks Carry Gas to Underground Tanks	Gas Comes from Oil Refinery	Gas Comes from Crude Oil at Refinery	Crude Oil Drilled from Large Underground Reserves	Heat and Pressure Turn Organisms into Oil	Fossils Buried Under Sea Hundreds of Millions of Years Ago

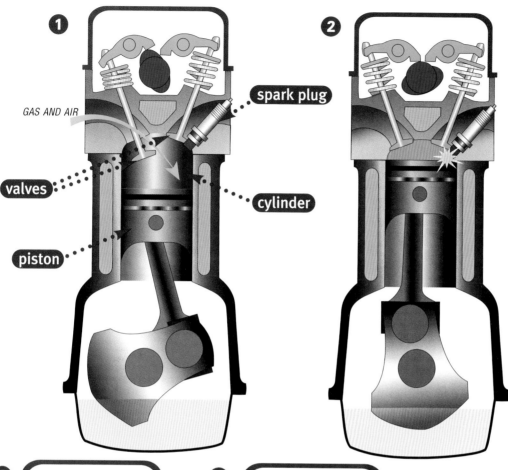

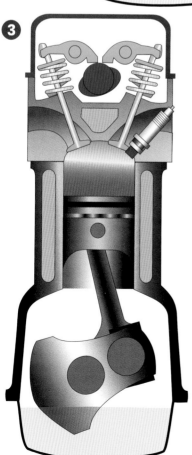

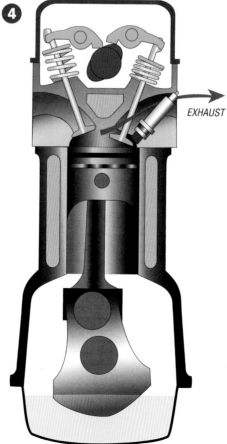

The Four-Stroke Combustion Cycle

1. *As the piston drops down, a valve opens. A mixture of gas and air enters the cylinder.*

2. *When the piston rises in the cylinder, it compresses the gas-and-air mixture, which the spark plug ignites.*

3. *The explosion forces the piston back down again.*

4. *As the piston rises, it forces the smoke from the explosion out through an open valve as exhaust.*

Start the Car

Side Step

TIMING IS EVERYTHING

If engines are exploding all the time, why do they seem to run so smoothly?

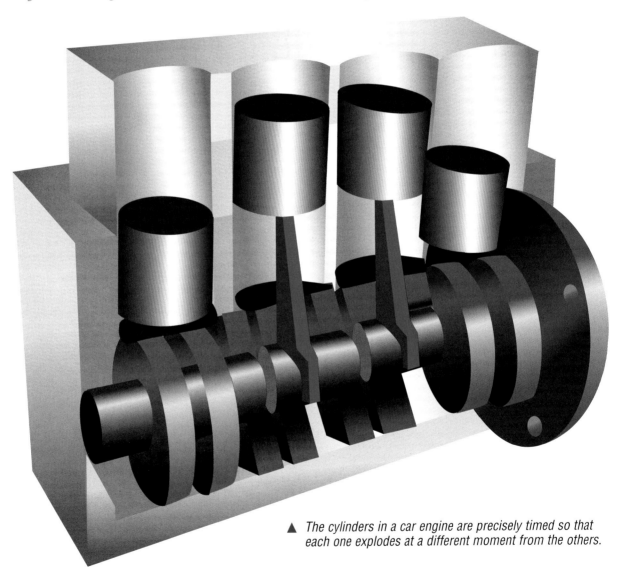

▲ The cylinders in a car engine are precisely timed so that each one explodes at a different moment from the others.

The average car engine has four cylinders, but they do not all work at the same time. When the engine is running, each cylinder goes through a four-part cycle:
1. Gas and air are sucked inside the cylinder.
2. The gas and air mixture is compressed by the piston.
3. The mixture is ignited by a spark and explodes.
4. Burnt gases are released.

For the engine to run smoothly, the cylinders explode in a sequence, one after the other. This sequence ensures that there is always enough power to keep the engine running smoothly. The spark plugs set off the explosions inside each cylinder, and a computerized system in the engine ensures that each spark plug receives an electrical current at a precise moment. The action happens repeatedly, several hundred times a minute!

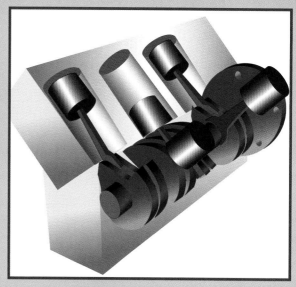

▲ Six cylinders are lined up in a V-shaped formation.

Form and Function

The more cylinders an engine has, the more smoothly and powerfully it runs. Depending on the size and design of the car, cylinders can be arranged in different ways.

If a car has only four cylinders, they are commonly set in a straight line. A compact sports car with six cylinders, however, is usually designed differently. If the six cylinders were lined up in a row, the engine would be too large to fit inside the car. Thus, most sports cars have a V-6 engine—six cylinders in a V formation—to pack a lot of power into a small space.

Start the Car with a Key → Key Turns Ignition On → Electricity Flows from Battery to Engine → Engine Runs on Explosions → Mixture of Gas and Air Explodes and Moves Pistons → **Driver Hits Gas Pedal to Squirt Gas in Engine** → Gas Added to Car at Gas Station

Driver Hits Gas Pedal to Squirt Gas in Engine

Where does the gas come from?

Gas comes to the engine from the fuel tank. Before it can make a powerful explosion inside the cylinder, it must be mixed with air. Older cars used a part called a carburetor to suck air into the engine. Many cars today use a fuel injection system. This system squirts the correct amount of air and gas directly into the cylinder each time. To guarantee accuracy, fuel injection is controlled by the computerized system in the car engine.

The car's gas supply is also connected to a pedal called the gas pedal, or accelerator. It is located on the car floor, to the right of the steering wheel. The driver feeds more gas to the engine by pressing one foot on the accelerator. This action controls how fast the car will move. More pressure on the gas pedal delivers more gas to the engine and makes the car move faster.

But how did the gas get into the gas tank?

| Trucks Carry Gas to Underground Tanks | Gas Comes from Oil Refinery | Gas Comes from Crude Oil at Refinery | Crude Oil Drilled from Large Underground Reserves | Heat and Pressure Turn Organisms into Oil | Fossils Buried Under Sea Hundreds of Millions of Years Ago |

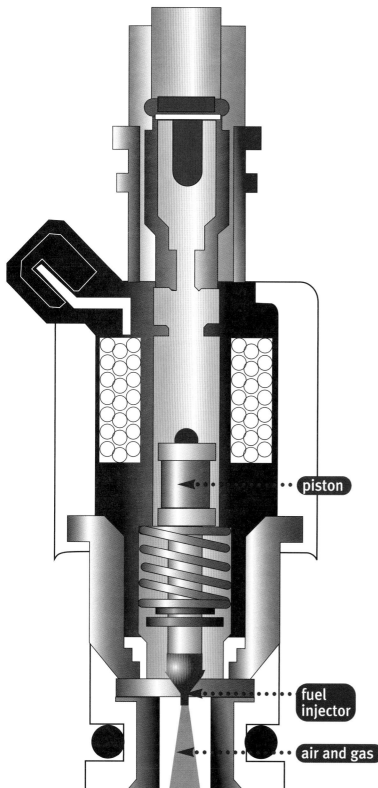

▲ Before cars, people used horses for transportation.

From a Horse, of Course

Today it is common to refer to a car's power as its horsepower. The origin of the word was, in fact, inspired by a horse. Long before cars were invented, people used horses for travel and heavy labor. In the 18th century, engineer James Watt coined the term horsepower as a measurement of force.

While in a coal mine, Watt observed how much coal a horse could lift in a minute and how far it could carry the load. He concluded that an average horse could lift 33,000 foot-pounds per minute. (Foot-pounds are found by multiplying the weight of the job by the distance it is moved.) One horsepower, therefore, is the force needed to lift 330 pounds of coal 100 feet in one minute, or 33 pounds of coal 1,000 feet in one minute.

◀ A fuel injector is a valve that is programmed to spray just the right amount of gas mixed with air directly into the car engine.

Start the Car 21

| Start the Car with a Key | Key Turns Ignition On | Electricity Flows from Battery to Engine | Engine Runs on Explosions | Mixture of Gas and Air Explodes and Moves Pistons | Driver Hits Gas Pedal to Squirt Gas in Engine | Gas Added to Car at Gas Station |

Gas Added to Car at Gas Station

How did the gas get into the gas tank?

Gas is added to the car's tank through a pump at a gas station. The customer or station attendant turns the pump on with the flick of a switch, which is located near the handle. He or she then puts the hose's tip, called a nozzle, into the car's gas tank. Gas immediately starts to flow into the tank. The pump keeps track of how many gallons of gas flow into the car and how much this will cost when the person has finished pumping gas.

Pipes connect the pumps to large tanks below ground. Each tank holds approximately 12,000 gallons of gas. The tanks are made of two layers of rust-free fiberglass to guard against gas leaks. The second layer can contain gas if the inner chamber cracks. Around the tanks are 2 feet of gravel. The gravel can absorb gas and protect the soil if both tank walls break.

The tanks are buried 12 to 14 feet underground for a few reasons. Gas begins to evaporate at a lower temperature than water. Storing it underground helps keep the inside temperature of the storage tank cool. It also reduces the risk of a fuel spill on the ground. A fire could start if a large quantity of gas spills and makes contact with oxygen. The heat of the fire could cause an above-ground tank to heat up and explode, which could easily endanger a whole neighborhood.

So how does the gas get to the gas station?

| Trucks Carry Gas to Underground Tanks | Gas Comes from Oil Refinery | Gas Comes from Crude Oil at Refinery | Crude Oil Drilled from Large Underground Reserves | Heat and Pressure Turn Organisms into Oil | Fossils Buried Under Sea Hundreds of Millions of Years Ago |

▲ A customer pumps gas by fitting the nozzle to the opening of her car's fuel tank.

Enough's Enough

Fortunately for gas pumpers with short attention spans, a gas pump will automatically shut off when a car's tank is filled. The pump can detect a full tank through a small pipe that runs from the nozzle to the pump handle. Air flows easily through this pipe while the tank is being filled. Yet once the gas in the tank rises high enough to block the hole in the nozzle, the air stops flowing. This signals to the pump that the tank is full, so it stops the flow of gas.

Side Step

MAKING THE GRADE

Are there different kinds of gas?

Gasoline comes in different grades, or octane ratings. Drivers can fill up their cars with low-octane, mid-grade, or high-octane gas. Low-octane gas usually costs less because it is cheaper to make, but it is rougher on a car engine than high-octane gas.

Low-octane gas ignites more easily than a high-octane gas, which over time will hurt a car's power. In fact, low-octane gas can ignite in the engine even without the spark from the spark plug because it is very flammable, especially when mixed with air and compressed. Each time the gas ignites because of compression, rather than from the spark,

▶ *Gas stations offer different grades of gas, all of them without lead, a harmful substance.*

it causes a pinging sound. This is commonly referred to as "knocking in the car engine," and it means the fuel is burning rapidly and unevenly. Over time, engines that use low-octane gas will still run, but not as powerfully.

▲ An old sign advertises leaded gas, which is now banned.

Fill It with Unleaded, Please

In the 1910s, people discovered that one cheap way to get better performance from a car engine was to add lead to gas. Lead keeps gas in an engine from igniting under compression. The trouble with lead is that it is also poisonous to humans and other living things. Besides worrisome lead-filled gas spills, the government found that dangerous amounts of lead were polluting the air as part of the car's exhaust fumes emitted from its tailpipe. As a result, leaded gas was banned in 1996.

| Start the Car with a Key | Key Turns Ignition On | Electricity Flows from Battery to Engine | Engine Runs on Explosions | Mixture of Gas and Air Explodes and Moves Pistons | Driver Hits Gas Pedal to Squirt Gas in Engine | Gas Added to Car at Gas Station |

Trucks Carry Gas to Underground Tanks

How does the gas get to the gas station?

Gas arrives at the gas station in large, tube-shaped tanks towed by trucks called fuel tankers. The tankers' tanks are made of thick aluminum or reinforced steel and are tightly sealed because gas, even in its liquid form, is flammable and extremely dangerous. Each tank is big enough to carry about 2,500 gallons of gas and has several compartments. If all of the gas was in one compartment, it would hinder the driver's control over the truck. If, for example, the driver made a sudden stop in traffic, the massive weight of the gas would be pushed toward the front of the truck and may make the truck jump forward. Breaking up the amount of gas into separate compartments reduces this weight worry.

The tanks' openings are equipped with valves and lockable covers so that fuel will not leak. Fuel tankers are inspected regularly to make sure they are fit to transport the fuel safely.

But where do the trucks pick up the gas?

| Trucks Carry Gas to Underground Tanks | Gas Comes from Oil Refinery | Gas Comes from Crude Oil at Refinery | Crude Oil Drilled from Large Underground Reserves | Heat and Pressure Turn Organisms into Oil | Fossils Buried Under Sea Hundreds of Millions of Years Ago |

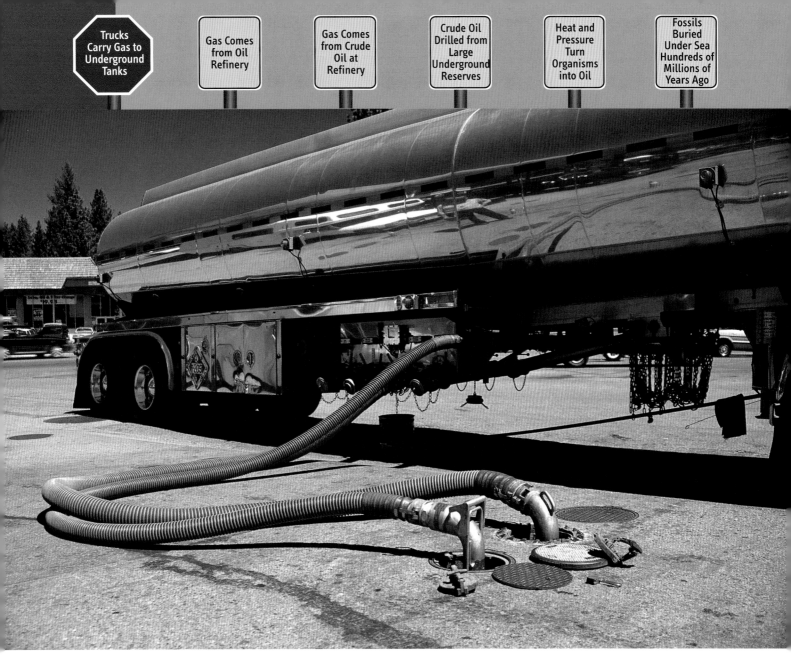

▲ A fuel tanker pumps gas into the underground storage tanks at a gas station.

▲ This 1922 photo shows an early gas station.

Open for Business

In the early days of automobiles, people had to get their gas from dry goods or hardware stores. They bought gas by the bucket and then poured it into their tank with a funnel. The growth of oil companies made it easier to purchase gas. Standard Oil Company (now Chevron) opened North America's first gas station in California in 1907. It provided gas through a 300-gallon tank and a garden hose.

| Start the Car with a Key | Key Turns Ignition On | Electricity Flows from Battery to Engine | Engine Runs on Explosions | Mixture of Gas and Air Explodes and Moves Pistons | Driver Hits Gas Pedal to Squirt Gas in Engine | Gas Added to Car at Gas Station |

Gas Comes from Oil Refinery

Where do the trucks pick up the gas?

Tanker trucks pick up gas from a fuel-loading facility. There gas is stored in enormous drum-shaped storage containers. These containers are heavily insulated to keep the temperature of the gas stable—particularly not too hot. When heated, gas turns into vapor. This is problematic because vapor takes up more space than liquid. Excessive heat could therefore cause the containers to explode under all the pressure.

The tanker truck connects to a storage container through a series of valves and hoses. These ensure that no air gets inside the truck's tank and that the correct pressure inside each container is maintained. Added air would increase the pressure inside the tank and cause the gas to become more flammable. Plus, once air and gas come under pressure in a small space, they could explode, just as they do in the car engine. Valves and hoses are inspected regularly at fuel-loading facilities to ensure air does not mix with gas prematurely.

Pipelines bring the gas to the fuel-loading facility. They are linked to a large industrial center called an oil refinery. Gas is made at the refinery.

How is gas made at the refinery?

| Trucks Carry Gas to Underground Tanks | **Gas Comes from Oil Refinery** | Gas Comes from Crude Oil at Refinery | Crude Oil Drilled from Large Underground Reserves | Heat and Pressure Turn Organisms into Oil | Fossils Buried Under Sea Hundreds of Millions of Years Ago |

▲ *Gas is stored in large cylindrical containers, ready for transport.*

Gas Guzzlers

People in the United States use about 131 billion gallons of gas per year. About 10 billion of these gallons are used to power cars.

Gas use is on the rise. By the year 2020, experts predict that North Americans will use about 1.37 billion gallons per day.

| Start the Car with a Key | Key Turns Ignition On | Electricity Flows from Battery to Engine | Engine Runs on Explosions | Mixture of Gas and Air Explodes and Moves Pistons | Driver Hits Gas Pedal to Squirt Gas in Engine | Gas Added to Car at Gas Station |

Gas Comes from Crude Oil at Refinery

How is gas made at the refinery?

At the refinery, gas is created from a thick black liquid called crude oil. Crude oil is made of molecules, or groups of atoms, called hydrocarbons. These hydrocarbons release a lot of energy when they are burned, so they are useful as fuel. The oil, however, contains many different chains of hydrocarbons of varied lengths. Before crude oil can be used as an energy source, the different kinds of hydrocarbons must be separated.

Separation of hydrocarbons is done through heating. Each hydrocarbon chain has its own boiling point. Chains with low boiling points turn to vapor first. The refinery collects this vapor and then cools it to return it to liquid form. The crude oil, meanwhile, continues to heat up. As other hydrocarbons approach their boiling points, they also turn to vapor and condense in separate containers.

Gas has a boiling point that ranges from 104°F to 752°F (40°C to 205°C). As crude oil reaches these temperatures, gas vapors collect in pipes and then cool down in separate containers. This forms liquid gas. By the time the crude oil has reached a maximum temperature of 1,112°F (600°C), hydrocarbons for at least eight different kinds of oil have been separated and collected.

After the heating process, the different oils are refined further through various chemical processes. The refinery removes impurities and adds other chemicals to make different oil by-products.

But where does crude oil come from?

| Trucks Carry Gas to Underground Tanks | Gas Comes from Oil Refinery | **Gas Comes from Crude Oil at Refinery** | Crude Oil Drilled from Large Underground Reserves | Heat and Pressure Turn Organisms into Oil | Fossils Buried Under Sea Hundreds of Millions of Years Ago |

▲ *Crude oil is heated inside these large, tubelike containers at an oil refinery.*

Harmful Helper

At the refinery, a chemical called MTBE (methanol tertiary butyl ether) is added to gas. This chemical helps boost gas's octane, which improves the performance of car engines. It also adds oxygen when the gas is burned and reduces the amount of harmful gases, such as carbon monoxide, that the engine produces.

MTBE, however, has a major flaw. Some studies suggest that it may cause cancer if it gets into the water supply. This could happen if an underground storage tank leaks and passes the chemical into the ground water. As a result of MTBE's health risk, a more expensive substitute, ethanol, may take the place of MTBE in the near future.

Side Step

WASTE NOT, WANT NOT

What else is produced at the refinery?

Oil refining creates more products besides gas. These by-products include other types of oil, such as heating oil, jet fuel, and lubricating oil. Car engines need lubricating oil to keep their parts running together smoothly.

Not every product that comes from crude oil is gas or oil. Tar and asphalt, for example, are some of the substances left behind at the end of the refining process. Furthermore, various chemicals produced by oil refining may be used to make plastic goods, such as containers and CD covers. Other products made from crude oil by-products include deodorants, crayons, bubble gum, and ink.

The process of oil refining also leaves behind a lot of waste, gases, and chemicals that are not useful for anything. Many of these materials are harmful to the environment if they are not disposed of properly.

One big problem with the oil refining process is the release of sulfur dioxide, a gas that contributes to sulfuric acid, into the atmosphere. This chemical mixes with water droplets to create acid rain, which damages trees, soil, and marine life and even erodes metallic structures. To protect the environment, most oil refineries now use filters and other cleaning systems to make the waste materials less harmful.

An Exhausting Problem

Pollution caused by car exhaust is a major environmental headache as well as a physical concern for living things. After an engine burns fuel, the waste products come out through the tailpipe in the back of the car. This exhaust is made of several gases, including carbon monoxide and nitrogen oxide. Carbon monoxide is poisonous, while nitrogen oxide causes polluted air called smog. Breathing in polluted air damages the lungs over time, especially in very young and elderly people.

In or near cities, smog may be visible even on a clear day. It appears as a layer of brownish gray smoke near the horizon. Most of this pollution comes from cars.

▲ Items such as crayons and plastic containers, and materials such as asphalt and ink, are all made from by-products of the oil refining process.

▲ Car exhaust is a major contributor to air pollution.

Start the Car 33

| Start the Car with a Key | Key Turns Ignition On | Electricity Flows from Battery to Engine | Engine Runs on Explosions | Mixture of Gas and Air Explodes and Moves Pistons | Driver Hits Gas Pedal to Squirt Gas in Engine | Gas Added to Car at Gas Station |

Crude Oil Drilled from Large Underground Reserves

Where does crude oil come from?

Huge natural reservoirs of crude oil lie hundreds of thousands of feet underground. In fact, another word for crude oil is petroleum, which means "oil from the earth."

To find out if these resources exist in a particular location, scientists learn about the geological history of the area and examine rock samples for evidence. If they find promising signs of oil, a drill is driven deep into the ground. People who do the drilling know they have found oil when the black fluid rises to the surface of the earth through the drill pipe. When oil is found, an oil rig, also called an oil well, is built on the site. The rig consists of the drill pipe, a rotating drill bit that can dig through rock and dirt, a mud pump, and a structure called a derrick that houses the drilling machinery.

The rig's main task is to dig through layers of solid rock to get to the oil reserves. Yet it also needs to get this rock out of the way to make a shaft down to the oil. At the end of the drill pipe, a drill bit grinds the rock into pieces. All this grinding produces a lot of heat. To cool down the drilling area, mud is pumped down through the pipe to the bit. When the mud rises back up to the surface through a shaft outside the drill pipe, it brings with it the ground up bits of rock. When the drill reaches the oil supply, crude oil flows up through the drill pipe to the surface.

So what is crude oil made of?

▲ An oil rig can tap into crude oil reserves by drilling through many layers of solid rock.

Reservations about Reserves

The United States has reserves of about 22.05 billion barrels of crude oil. The supply is underground and just needs to be drilled up to the surface. This amount of reserve oil may sound like a lot, but it is not an endless supply. The United States pays a great deal of money to import half of the oil it needs from other crude oil producing nations, such as Saudi Arabia.

If these oil exporters raise crude oil prices too high or stop exporting oil, the U.S. government can draw oil from its Strategic Petroleum Reserve. This is an additional supply of 570 million barrels of crude oil that is stored in underground salt caves in Louisiana. If the United States stops buying oil from other countries and instead draws oil only from this reserve, it would run out after 60 days.

Start the Car

| Start the Car with a Key | Key Turns Ignition On | Electricity Flows from Battery to Engine | Engine Runs on Explosions | Mixture of Gas and Air Explodes and Moves Pistons | Driver Hits Gas Pedal to Squirt Gas in Engine | Gas Added to Car at Gas Station |

Heat and Pressure Turn Organisms into Oil

What is crude oil made of?

Crude oil is known as a fossil fuel, which means that it is made of the fossils of ancient plants and animals. After these things died, their remains decayed and formed a layer of dead organic matter. Hundreds of millions of years of tremendous heat and pressure under the earth turned this layer into oil. Over time, the oil was trapped under more layers of rock.

All living things contain a natural element called carbon, which is released into the air when an organism dies and its remains decay. In fossil fuels, however, the carbon is contained in the organic remains. Over time, it is stored in the chains of hydrocarbons. The carbon in fossil fuels is released into the air only after the crude oil is brought to the surface, refined into petroleum products, and burned as fuel for energy.

But how did the fossils get deep underground?

Finding Other Fossil Fuels

Fossil fuels exist in other forms underground. Coal, for example, is a fossil fuel in solid form. Though fossil fuels are constantly being produced, they take many millions of years to form. The coal used today formed from ancient trees that collapsed into swamps and were buried in layers of mud. Like oil, coal is buried deep underground. Because it is solid, coal cannot be drilled. The only way to get it is to dig into the earth and scoop it up with heavy equipment.

Natural gas is another fossil fuel trapped in underground deposits. Most of it is in the form of methane gas. Natural gas is collected by pipes drilled into the ground. The gases can then flow up naturally through the pipes as ready fuel for heating and industry. Unlike oil and coal, natural gas burns cleanly with few pollutants.

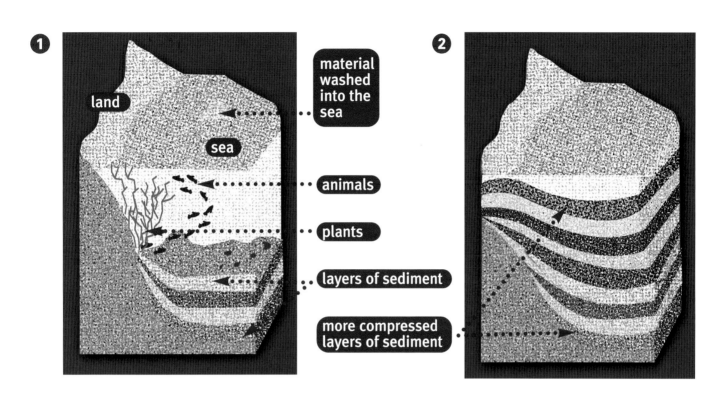

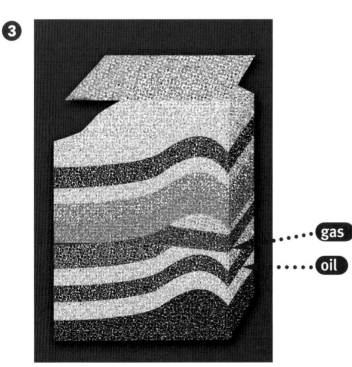

How Petroleum Is Made

1. Plants and animals die, and their remains settle on the ocean floor with other materials washed into the sea, such as sand. Layers of sediment begin to form.

2. Over time, this organic matter is buried by more compressed layers that harden into rock.

3. Heat and pressure turn the organic matter into gas and oil, which are trapped beneath the layers.

| Start the Car with a Key | Key Turns Ignition On | Electricity Flows from Battery to Engine | Engine Runs on Explosions | Mixture of Gas and Air Explodes and Moves Pistons | Driver Hits Gas Pedal to Squirt Gas in Engine | Gas Added to Car at Gas Station |

Fossils Buried Under Sea Hundreds of Millions of Years Ago

How did the fossils get deep underground?

Over long periods of time, deposits of earth materials covered and compressed the remains of living things. The process to make the crude oil that is used today first began 400 million years ago in a warm sea that was home to many tiny plants and animals. As these organisms died, their remains settled on the sea bottom. They were buried under deposits of mud and sand called sediment.

Gradually, about 50 to 100 million years ago, the sediment hardened into rock. This trapped the plant and animal remains underground. The weight of the sea and the layers of rock weighed down on the organic matter. It is this pressure, plus heat from the center of the earth, that gradually changed the remains to oil.

Because crude oil takes hundreds of millions of years to form, dinosaurs ruled the earth by the time the organic matter started turning into oil. Today, fossil fuels are still forming, but very slowly. Animal remains on the bottom of oceans today should become crude oil 400 million years from now. People are using fossil fuels much faster than they can develop, so one day soon fossil fuels may very well run out. This is why fossil fuels are called a nonrenewable source of energy.

| Trucks Carry Gas to Underground Tanks | Gas Comes from Oil Refinery | Gas Comes from Crude Oil at Refinery | Crude Oil Drilled from Large Underground Reserves | Heat and Pressure Turn Organisms into Oil | Fossils Buried Under Sea Hundreds of Millions of Years Ago |

▲ Plant and animal remains on the ocean floor today could become crude oil 400 million years from now.

Drilling for Gold

▲ Oil wells like this one are common in Texas.

The oil industry got its start on August 27, 1859. On that date in Titusville, Pennsylvania, former railroad conductor Edwin Laurentine Drake drilled 69 feet into the ground with an old steam engine. His drill struck oil, and Drake built an oil well on the site. Soon other oil seekers flocked to the area and drilled their own wells. Eventually, they drove Drake out of business. In the years following Drake's discovery, oil manufacturers began refining crude oil for many purposes, particularly for lighting fuel such as kerosene. The invention of the car engine was just around the corner.

Start the Car

The Big Picture

Learn how a car starts and what makes it go.

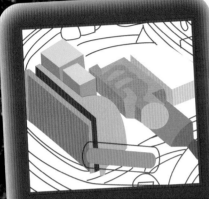

13 Start the Car with a Key
A turn of the key turns the car on and starts the engine.
(pages 6–7)

11 Electricity Flows from Battery to Engine
The starter motor draws electricity from the battery to give the engine its first crank. This action makes the engine run.
(pages 10–11)

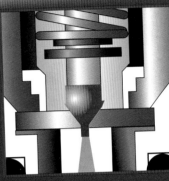

8 Driver Hits Gas Pedal to Squirt Gas in Engine
Once the car is in motion, the driver can control the speed by feeding gas from the gas tank to the engine. More gas makes the pistons move up and down more quickly, which causes the car to speed up.
(pages 20–21)

12 Key Turns Ignition On
The turned key in the ignition supplies a current of electricity to the engine's starter motor.
(pages 8–9)

10 Engine Runs on Explosions
Sparks inside the engine set off explosions, which push pistons inside the engine's cylinders up and down. The pistons connect to a shaft that makes the car wheels turn.
(pages 14–15)

9 Mixture of Gas and Air Explodes and Moves Pistons
The pistons compress a mixture of gas and air at the top of the cylinder. The spark ignites this flammable mixture and causes an explosion, which forces the pistons back down.
(pages 16–17)

⑦ Gas Added to Car at Gas Station

Gas stations store gas in large underground tanks. The gas is pumped into the car's gas tank through hoses.
(pages 22–23)

⑤ Gas Comes from Oil Refinery

Gas is loaded onto the trucks at fuel loading facilities. It reaches the loading facility through a pipeline from an oil refinery.
(pages 28–29)

③ Crude Oil Drilled from Large Underground Reserves

Crude oil comes out of oil rigs, which are drilled deep into the ground.
(pages 34–35)

⑥ Trucks Carry Gas to Underground Tanks

Trucks called fuel tankers bring gas to gas stations.
(pages 26–27)

④ Gas Comes from Crude Oil at Refinery

The oil refinery takes crude oil and treats it in various ways to produce gas.
(pages 30–31)

② Heat and Pressure Turn Organisms into Oil

Oil reserves were once the remains of living things. Heat and pressure changed this matter into oil.
(pages 36–37)

① Fossils Buried Under Sea Hundreds of Millions of Years Ago

Organic remains collected at the bottom of an ancient ocean. They were buried in mud, which turned to rock over millions of years.
(pages 38–39)

Start the Car

Facts and Figures
Car Mall

Solar Power Studies

For decades, scientists have attempted to perfect a solar-powered car that can run on the natural energy of the sun and create no pollution. Though too expensive to mass-produce at this time, the solar-powered cars they have developed use large solar panels to convert sunlight to energy. This energy charges a car's battery, which feeds the car motor. In August 2000, a group of college students drove a solar-powered car across Canada on just 1,000 watts of power. That is the same amount of energy needed to operate a toaster! The journey took about four weeks at an average speed of 50 miles per hour.

A Drive Through Time
Many different inventions contributed to the development of the car.

1680	1769	1860	1876
Physicist Christiaan Huygens of Holland designs the first internal combustion engine based on principles that will later power the car engine.	French engineer Nicolas-Joseph Cugnot invents the first automobile, which is powered by a steam engine.	Jean-Joseph-Étienne Lenoir, a Belgian engineer, patents an internal combustion engine that uses coal gas as fuel. Three years later, Lenoir improves his invention so it burns petroleum, and he uses it to drive a wagon 50 miles.	Nikolaus August Otto of Germany invents the Otto Cycle, a four-stroke internal combustion engine.

Time for a Change

The first cars looked exactly like carriages without the gear needed to harness a team of horses. Soon automakers realized they could reduce the size of the wheels and bring the cars lower to the ground because passengers no longer had to worry about getting hit by mud and stones from horses' hooves.

Microchip Miracles

Thanks to computers, cars today are becoming more sophisticated. High-tech electronics are behind the car's fuel injection system to make sure the mixture of gas and air is just right. Computerized systems also make sure the spark plugs in the engine ignite at precise moments. Some cars' computerized systems even control maximum speed and indicate how quickly the engine burns gas.

A Price to Pay

Here is how gas prices have changed over a recent ten-year period.

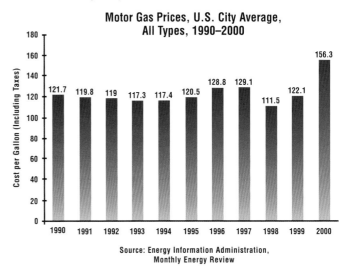

Motor Gas Prices, U.S. City Average, All Types, 1990–2000

Year	Cost per Gallon (Including Taxes)
1990	121.7
1991	119.8
1992	119
1993	117.3
1994	117.4
1995	120.5
1996	128.8
1997	129.1
1998	111.5
1999	122.1
2000	156.3

Source: Energy Information Administration, Monthly Energy Review

1885
German engineer Carl Benz builds the first automobile powered by the four-stroke engine. It has three wheels.

Germans Gottlieb Daimler and Wilhelm Maybach improve on Otto's engine and develop the forerunner for the modern car engine.

1886
Daimler attaches his engine to a stagecoach and builds the first four-wheeled automobile.

1900
Carl Benz's company, Benz & Co., becomes the largest maker of automobiles.

1908
The Ford Motor Company manufactures the first Model T Fords at its plant in Highland Park, Michigan. The company's assembly line system produces automobiles at a cheaper cost and in large numbers. This enables cars to be made more widely available. The age of the automobile has begun.

Paying at the Pump

People buy gas by the gallon. Gas prices go up and down all the time because they are set by several factors. The main factor is the price of crude oil. After that, gas suppliers add the refinery's costs to make gas from crude oil, plus money that goes to service providers such as the fuel tanker companies and gas stations. Finally, taxes are added. State and federal taxes make up about 36 percent of the cost of a gallon of gas.

Start the Car

Wonders and Words

Questions and Answers

Q: *How fast can the average car go?*

A: The average car has the ability to travel at a speed of 100 to 120 miles per hour. Fast cars have more powerful engines than slower cars. These engines usually have more cylinders to provide the extra power needed to reach high speeds quickly.

Q: *How are trucks different from cars?*

A: Trucks have larger engines with much larger cylinders. A larger cylinder means more space for the gas and air mixture as well as a larger explosion. Trucks need this extra power to carry heavy loads.

Q: *When will the world's supply of crude oil run out?*

A: Experts predict that the known reserves of petroleum will last until about the year 2040. Supplies could run out sooner, however, as the demand for oil is steadily increasing around the world. Alternatively, advanced technology may make oil uses more efficient, so supplies could last longer. Companies and governments are exploring other sources of energy so that the world will not be so dependent on oil.

Glossary

Condense: To change from a gas to a liquid through cooling

Crankshaft: A shaft turned by the pistons at one end that turns the car's wheels on the other end

Crude Oil: A thick, black substance found underground, made from ancient remains of plants and animals

Cylinder: Part of a car engine that contains the piston and covers the explosion of gas and air

Derrick: A structure built over an oil rig to house tools and drilling equipment

Evaporate: To change from a liquid to a gas through heat

Exhaust: The burnt gases released by a car engine

Fossil: A trace of a long-dead plant or animal

Fuel Tanker: A truck designed to carry gas from one place to another

Gas (Gasoline): A product extracted from crude oil, used as fuel for cars

Hydrocarbon: An energy-bearing molecule found in crude oil

Internal Combustion Engine: An engine that gets its force from a controlled explosion inside

Oil Refinery: An industrial complex where gas and other products are extracted from crude oil

Oil Rig: Also oil well. A mechanism that gets crude oil out of the ground

Organic Matter: Material that is, or used to be, alive

Petroleum: Another word for crude oil that literally means "oil from the earth"

Piston: Part of a car engine that moves up and down inside the cylinder and turns the crankshaft

Car Talk

Advertisements for cars list various features and selling points. Here is what they mean:

V-6: Describes the engine (in this case, a six-cylinder engine with cylinders in a V-formation)

All-wheel drive: The driving mechanism connects to all four wheels, not just to the front or rear wheels, in order to provide better control in difficult road conditions

Independent rear suspension: The rear wheels each have their own suspension system, which makes the car easier and safer to drive

Power steering: A motor connected to the steering mechanism that makes it easier to turn the car

Liter size: Refers to the size of the cylinders in the engine (usually a bigger cylinder means a more powerful engine)

Turbo-charged: Exhaust gases drive a turbine that pumps the gas and air mixture into the engine, which increases its acceleration

Anti-lock brakes: Brakes that are controlled by a computer to bring the car to a sudden stop without locking the wheels

Index

Acid rain, 32
Air, 16 17, 19, 20, 21, 23, 28, 40
   See also oxygen
Animals, 36, 37, 38, 39, 41
Asphalt, 32, 33

Benz, Carl, 43

California, 26
Car
  accelerator, 7, 20, 40
  battery, 7, 10, 12, 13, 40, 42
  carburetor, 20
  coil, 8, 10
  crankshaft, 14, 16, 17, 40, 45
  early, 11, 20, 42, 43
  electric, 12, 13
  how they work, 6, 7, 10, 22, 14, 15, 16, 18, 40, 41
  ignition, 6, 8, 9, 10, 11, 40
  keys, 6, 8, 9, 40
  parts of, 6, 7
  solar-powered, 42
  starter motor, 8, 11, 40
  wheels, 6, 7, 17, 45

Car engine, 7, 8, 10, 13, 14, 19, 20, 24, 25, 31, 40, 42, 43
  cylinder, 14, 16, 17, 18, 19, 20, 45
  parts of, 14, 15, 16, 27
  piston, 14, 16, 17, 19, 40, 45
  spark plug, 14, 15, 19, 41
Car theft, 8, 9
Carbon, 36
Carbon monoxide, 33
Coal, 36
Computer system, in car, 19, 20, 43
Cugnot, Nicolas-Joseph, 42

Daimler, Gottlieb, 43
Dinosaurs, 38
Drake, Edwin Laurentine, 39

Electrical power in car, 7, 8, 10, 12, 13, 19
Electricity, 8, 9, 10, 19
Energy, 12, 14, 30, 36, 38, 42, 44, 45
Ethanol, 31
Exhaust, 16, 19, 25, 45

Ford Motor Company, 43
Fossils, 36, 37, 38, 41, 45
Fuel, 10, 32
   See also gas
  fossil, 36, 37, 38, 41
Fuel cells, 12
Fuel injection system, 20, 21, 41, 43
Fuel-loading facility, 28, 41
Fuel storage tanks, 22, 26, 27, 28, 29, 31, 41
Fuel tank, car, 7, 20, 22, 23
Fuel tankers, 26, 28, 41, 43, 45

Gas pedal, See car, accelerator
Gas station, 22, 24, 26, 27, 41, 43
Gas, 7, 12, 13, 16, 17, 19, 22, 24, 26, 28, 29, 30, 31, 40, 43
  consumption of, 29, 44
  dangers of, 22, 25, 26, 28
  types of, 24, 25, 31

Heat, 14, 15, 22, 28, 30, 31, 34, 36, 37, 41
Highland Park (Michigan), 43

Horsepower, 21, 42
Huygens, Christiaan, 42
Hydrocarbons, 30, 36, 45
Hydrogen, 12

Kerosene, 39

Lead, 24, 25
Lenoir, Jean-Joseph-Étienne, 42
Louisiana, 35

Maybach, Wilhelm, 43
Methane, 36
MTBE (methanol tertiary butyl ether), 31

Natural gas, 36
Nitrogen oxide, 33

Ocean, 36, 38, 39, 41
Oil
  crude, 30, 32, 34, 35, 36, 38, 39, 41, 43, 44, 45

drilling for, 34, 39, 41
lubricating, 32
origins of, 36, 37, 38, 41, 45
pipeline, 28
refining of, 28, 29, 32, 41, 43, 45
rig, 34, 35, 39, 41, 45,
Oil well, See oil rig
Organic matter, 36, 37, 38, 41, 45
Otto, Nikolaus August, 42, 43
Oxygen, 12, 22

Petroleum, See oil, crude
Plants, 36, 37, 38, 39, 41
Pollution, 12, 25, 32, 33, 36, 42

Race car, 16
Rock layers, 34, 35, 36, 37, 38, 41

Saudi Arabia, 35

Sediments, 37, 38, 41, 45
Smog, 33
Spark, 9, 10, 14, 15, 22, 25, 40
Speed, 16, 20, 40, 43, 44
Sports car, 19
Standard Oil Company, 26
Strategic Petroleum Reserve, 35
Sulfur dioxide, 32
Sulfuric acid, 312

Tanker trucks, See fuel tankers
Tar, 32
Titusville (Pennsylvania), 39
Trees, 36
Trucks, 26, 44

Watt, James, 21

Credits:

Produced by: J.A. Ball Associates, Inc.
Jacqueline Ball, Justine Ciovacco, Andrew Willett
Daniel H. Franck, Ph.D., Science Consultant

Art Direction, Design, and Production:
designlabnyc
Todd Cooper, Sonia Gauba

Writer: Lelia Mander

Cover: Bruce Glassman: ignition; Courtesy of Toyota:engine; Brooke Fasani: pumping gas; PhotoDisc, Inc.: tanker; Ablestock: oil refinery.

Interior: Ablestock/Hemera: p.3 traffic, pp.14–15 spark plug (large), p.29 traffic, p.31 oil refinery, p.33 asphalt, crayons, plastic dishes, p.35 gasoline can, p.40, 48 winding road (background), pp.44–45 traffic; ArtToday: p.6 battery, p.7 fuel tank, accelerator, p.16 race cars, p.36 coal mine; Photospin: p.7 engine, p.15 spark plug (small), p.23 fuel pump, p.25 leaded gasoline sign, p.29 gasoline storage, p.33 bottle of ink, car exhaust, p.35 drilling for oil, p.42 moving car, p.43 computer monitor, p.44 truck; Bruce Glassman: p.7 ignition, p.9 ignition; Sonia Gauba: p.9 hotwire, p.17 diagram of combustion cycle, p.18 diagram of four cylinder engine, diagram of six cylinder engine, p.21 diagram of fuel injection, p.37 diagram of fossil fuels; Courtesy of Toyota: p.11 engine, p.13 hybrid vehicle; Library of Congress: p.11 early cars, p.21 horse-drawn carriage, p.27 early gas station, p.39 old oil well, p.42 old car; Courtesy of DaimlerChrysler Corporation: p.12 GEM electric vehicle; Brooke Fasani Photography: p.23 woman with pink scarf pumping gas, p.25 gas pumps, p43 pumping gas; PhotoDisc, Inc.: p.27 tanker, p.39 ocean floor

For More Information

Farndon, John. *How the Earth Works.* Pleasantville, NY: Reader's Digest, 1992.

Lord, Harvey G. *Car Care for Kids and Former Kids.* New York: Atheneum, 1984.

Macaulay, David. *The Way Things Work.* Boston: Houghton Mifflin, 1988.

Parker, Steve. *53½ Things That Changed the World* Brookfield, CT: The Millbrook Press, 1992.

Sutton, Richard. *Eyewitness Books: Car.* New York: Alfred A. Knopf, 1990.

Tahta, Sophy. *What Makes a Car Go.* Boston: EDC, 1994.